AF207094

URANUS

by Heather C. Morris

BrightPoint Press

San Diego, CA

© 2026 BrightPoint Press
an imprint of ReferencePoint Press, Inc.
Printed in the United States

For more information, contact:
BrightPoint Press
PO Box 27779
San Diego, CA 92198
www.BrightPointPress.com

LIBRARY OF CONGRESS CATALOGING-IN-PUBLICATION DATA

Name: Morris, Heather C., author.
Title: Uranus / by Heather C. Morris.
Description: San Diego, CA: ReferencePoint Press, 2026 | Series: The planets of our solar system | Audience: Grade 7 to 9 | Includes bibliographical references and index.
Identifiers: ISBN: 9781678211745 (hardcover) | ISBN: 9781678211752 (eBook)
The complete Library of Congress record is available at www.loc.gov.

CONTENTS

AT A GLANCE

- Uranus is the seventh planet from the Sun. It is the third-largest planet in the solar system.

- Uranus is one of four giant planets in the solar system. Uranus's atmosphere is made of a combination of gases and frozen material.

- It takes 84 years on Earth for Uranus to complete one orbit around the Sun.

- People in ancient times could see Uranus in the night sky, but many thought it was a star. The invention of the telescope led people to discover that it was a planet.

- Space telescopes orbiting Earth provide better views of Uranus than ground-based telescopes. Scientists have learned about Uranus using several space telescopes.

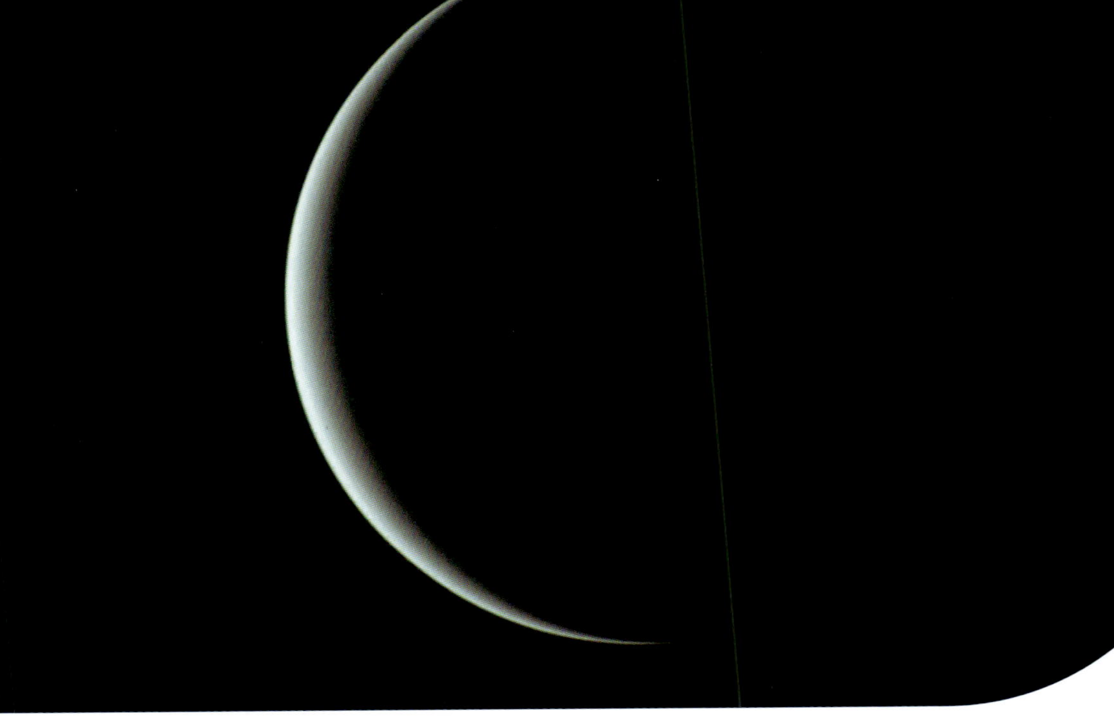

- Only one spacecraft has ever visited Uranus. *Voyager 2* flew by the planet in 1986.

- Uranus has twenty-eight known moons and thirteen rings. All of Uranus's moons are named after characters in works of literature by William Shakespeare or Alexander Pope.

- Uranus's moons can be divided into three groups. These are the inner moons, irregular moons, and major moons.

SEEING URANUS IN A NEW LIGHT

n 2023, the new James Webb Space Telescope (JWST) pointed its camera toward the distant planet Uranus. It was the first time this telescope had taken photos of the icy world. The Hubble Space Telescope (HST) had snapped a few photos since it was launched in 1990. And the spacecraft *Voyager 2* had photographed the planet as it flew by in 1986. Both HST and *Voyager 2*

The James Webb Space Telescope uses mirrors to direct light from space onto sensitive scientific instruments.

NASA
God
SPACE F

In JWST's images of Uranus, the planet's rings appear much brighter than they do to the human eye.

had used cameras that detect visible light. This is the light that people can see.

But JWST is different. It has special cameras. These cameras can detect light that people cannot see with their eyes.

Astronomers wondered what JWST would show them about Uranus.

Soon, the photos were available. In them, Uranus and its rings glowed. Scientists could see Uranus's faint inner and outer rings. The photos also showed that Uranus's **atmosphere** is active. Clouds covered the north pole. Storms swirled around the middle of the planet.

A few months later, JWST snapped more photos. These pictures captured Uranus's large moons. The clouds at the north pole were brighter in these new images. And the storms elsewhere on Uranus had moved or changed.

JWST helped scientists study how weather changes on Uranus. This information will be used in planning future missions to the distant planet. Future spacecraft will join JWST in helping people learn about this unusual world.

A DISTANT WORLD

Uranus orbits the Sun at a great distance. People have seen Uranus in the sky since

JWST captured this image of Uranus and some of its moons, as well as several background galaxies.

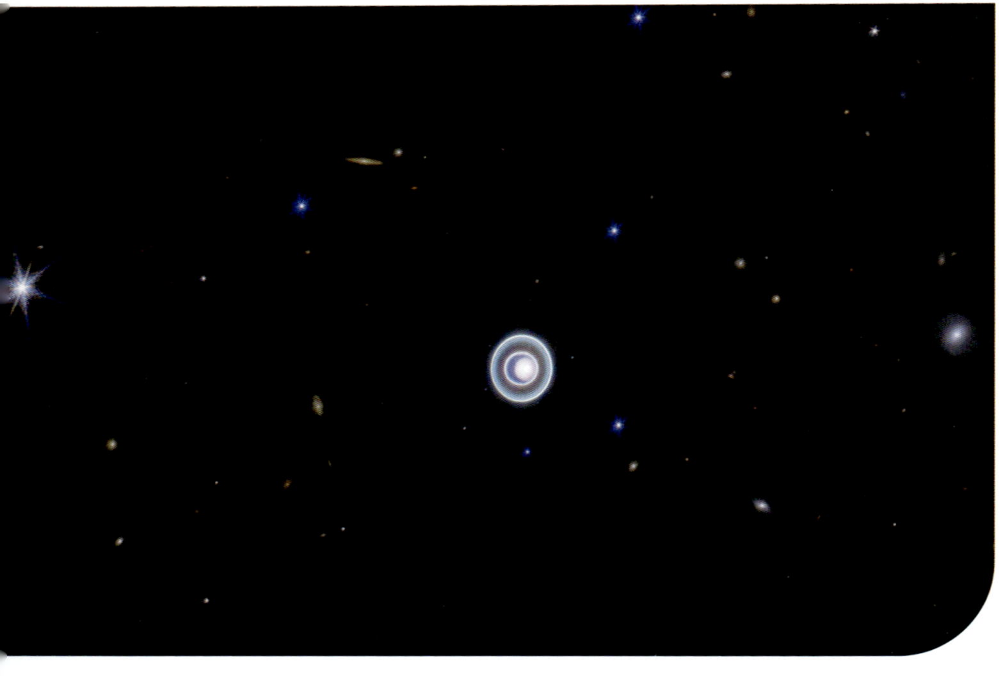

ancient times. But it is very dim. People once thought it was a star or something else. Then telescopes revealed that Uranus was a planet. Because the planet is so far from Earth, it is difficult to send a spacecraft there. The journey takes many years.

Despite these challenges, people have learned a lot about Uranus. *Voyager 2* took up-close photos of the planet. And telescopes such as HST and JWST have uncovered more of its mysteries. But many questions remain unanswered. Scientists continue to study Uranus.

CHAPTER ONE

AN ICE GIANT

Uranus is a cold, dark planet. It is about 1.8 billion miles (2.9 billion km) from the Sun on average. This distance is so great that it takes sunlight about 2 hours and 40 minutes to reach the planet. When it arrives, the light is weak and dim. Neptune is the only planet farther away from the Sun than Uranus. But Uranus holds the record for the lowest temperature in the Solar System. Temperatures have reached as low

Uranus's unique features fascinate astronomers.

Neptune is slightly smaller than Uranus.

as −372°F (−224°C). The planet's average temperature is a little warmer at −320°F (−195°C). Scientists are studying why Uranus is so cold.

Uranus is the seventh planet from the Sun. It is the third-largest planet by diameter in the solar system. Uranus is about 31,763 miles (51,118 km) across. This size makes Uranus about four times larger than Earth. Its mass is about fourteen times the mass of Earth. But its volume is much larger. About sixty-three Earths would fit inside Uranus.

Uranus and its neighbor, Neptune, are twins. They are made of similar materials. They are almost the same size. And they are similar in color. Scientists think both planets may have formed closer to the Sun. Then they slowly drifted farther out.

There are twenty-eight known moons and thirteen rings orbiting Uranus. The rings are divided into two sets. There are eleven inner rings. They are mostly narrow, gray, and dark. There are two outer rings. They are wide and very faint.

ICE IN THE ATMOSPHERE

Uranus is one of the four giant planets in the solar system. The others are Jupiter, Saturn, and Neptune. A giant planet is a planet much larger than Earth. These planets usually do not have solid surfaces. Such planets are unlike terrestrial planets such as Earth. Terrestrial planets have rocky surfaces.

Uranus is mostly made of methane, water, and ammonia. The methane in Uranus's atmosphere absorbs red light. That means the planet reflects little of this light. This causes Uranus to appear blue-green.

Uranus is one of the least dense planets in the solar system. But Uranus is different from Saturn and Jupiter. It is so cold that

Jupiter is the most massive planet in the solar system.

icy slush swirls among the gases in its atmosphere. This is why scientists call Uranus an ice giant. Neptune also falls into this category.

The winds on Uranus are strong. They can reach speeds of about 560 miles per hour (900 kmh). That is more than twice as fast as the fastest wind speed recorded on Earth. At the **equator**, winds race around the planet against the direction of Uranus's spin.

Even though it has strong winds, Uranus does not have many storms.

Special telescopes can see details in Uranus's atmosphere that regular cameras cannot.

Scientists do not know why Uranus has fewer storms than the other giant planets. When *Voyager 2* visited Uranus, scientists were surprised by how calm the planet looked. But storms do occur on Uranus. Since 1986, scientists have watched the planet through telescopes. They have found shifting clouds as Uranus changes seasons.

Uranus spins quickly. One day on Uranus only lasts about 17 hours. But 1 year on Uranus is a very long time. It takes about 84 Earth years for Uranus to travel around the Sun once.

A MYSTERIOUS PLANET

Uranus has many mysterious features that interest scientists. For example, its **axis of rotation** is tilted. The planet lies on its side. No one knows exactly why Uranus tipped over.

The way Uranus turns is another mystery. Uranus rotates in the opposite direction of most of the other planets. Venus is the only other planet that spins in this way.

Uranus's magnetic field is also different. A magnetic field is a region around a planet in which magnetic forces occur. Many planets

Collision?

Scientists wonder whether something struck Uranus in the distant past. When the solar system was young, large rocks struck planets more often than they do now. Some of these rocks were the size of moons. One of them might have slammed into Uranus. Some scientists argue that this could have tilted the planet.

Uranus's extreme tilt makes it stand out among the planets of the solar system.

have a magnetic field. But Uranus's is
unique in some ways. Its magnetic field
does not line up with its axis. And it is offset
from the center of the planet. Amy Simon is
a scientist at the National Aeronautics and
Space Administration (NASA). She studies
the atmospheres of planets. She says,

Unlike Uranus, Saturn has storms that are easily visible to the human eye.

"[Uranus has] a very complex, weird magnetic field."[1]

Scientists are not sure what Uranus is like beneath its clouds. They suspect that it has a solid core. But they are not sure about what lies between the tops of Uranus's clouds and the planet's core. "Does Uranus have an ocean layer and then an atmospheric layer? . . . We just don't know," says astronomer Heidi Hammel.[2]

Uranus's strong winds are another puzzle. Scientists think that heat inside giant planets helps fuel fast winds. But Uranus is a very cold planet. Something else might be churning up Uranus's winds.

CHAPTER TWO

DISCOVERING URANUS

Uranus has puzzled astronomers for thousands of years. This planet is visible in the night sky without a telescope. But it is very dim. Because it is far away from the Sun, Uranus has a slow orbit. This means it moves slowly in the sky. The other visible planets are much closer. They move relatively quickly across the sky. This is why many people who saw Uranus in ancient times did not realize it was a planet.

People have studied the night sky since before written history.

John Flamsteed was the first director of the Royal Observatory, Greenwich. This institution was founded in 1675.

The astronomer Hipparchus lived in ancient Greece. He made a list of stars. Today, scientists can still find most of the stars on his list. But one of the stars is mysterious. It is not in the place that

Hipparchus said it was. Some researchers suggest that this star was actually Uranus.

In 1690, English astronomer John Flamsteed spotted Uranus. He observed the planet several more times. But he thought it was a star. He found it in the **constellation** Taurus. So, he called it 34 Tauri.

French astronomer Pierre Charles Le Monnier watched Uranus through a telescope 60 years later. He saw it on many different nights. Like earlier astronomers, Le Monnier thought Uranus was a star.

In March 1781, English astronomer William Herschel also saw Uranus. He thought it was a **comet**. But other astronomers suspected that Uranus was not a comet. Swedish astronomer Anders

This replica of one of Herschel's telescopes shows the enormous size of the original instrument he used.

Johan Lexell was the first to calculate Uranus's orbit. Other astronomers pointed out that its orbit was nearly circular like those of most planets. By 1783, even Herschel agreed that he had found a new planet. "It appears that the new star . . . is a Primary Planet of our Solar System," he wrote in a letter.[3]

After Herschel announced his discovery, Le Monnier sorted through his papers. He realized he had made a mistake. "Le Monnier examined his own records and found that he had seen Uranus nine times before its discovery without recognizing

William Herschel was born on November 15, 1738.

that it was not a star," writes science author Mark Littmann.[4]

The King of England at this time was King George III. He was thrilled at the idea of a new planet. He invited Herschel to move to a place near a royal castle. George III wanted the royal family to have access to Herschel's telescopes.

NAMING THE NEW PLANET

At first, no one knew what to name the planet. People had many different ideas. An astronomer named Jérôme Lalande wanted to name the planet after Herschel. Others proposed the names Astraea, Cybele, Minerva, and Neptune.

In 1782, German astronomer Johann Elert Bode suggested the name Uranus.

This was the Latin name of the Greek god of the sky. Bode argued that the new planet should be named like the other planets. This meant its name should come from ancient Greek and Roman stories. Many people liked Bode's suggestion. People began to use the name Uranus.

However, not every culture uses this name today. Uranus goes by different

George's Star

At first, William Herschel could not decide what to name the new planet. He called it Georgium Sidus. That is Latin for George's Star. He named it after the King of England. This name was not popular in other countries. Most people called the planet Uranus instead. Over time, people in the United Kingdom began to use the name Uranus as well.

names around the world. In China and some other Asian countries, people call this planet the Sky King Star. One of its names in Thai means Star of Death. And in Hawaiian, Uranus is called *Hele'ekala*. This is Hawaiian for the name Herschel.

MORE DISCOVERIES

Astronomers continued to watch Uranus after it was recognized as a planet. Herschel later found two of Uranus's moons. He named them Oberon and Titania. Over the next 100 years, people discovered three more moons. But astronomers still did not know much about the planet itself.

On March 10, 1977, a group of astronomers watched Uranus as it passed in front of a distant star. They noticed that

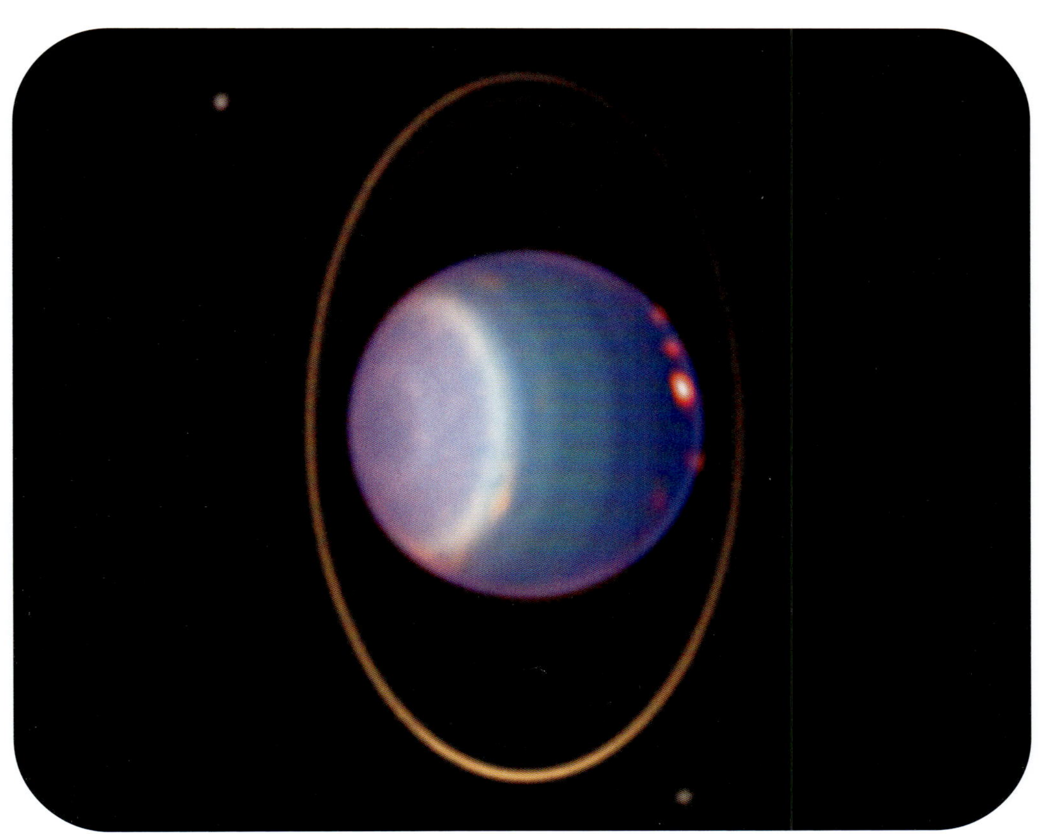

Uranus's rings are named after numbers and Greek letters.

the star's light dimmed several times before and after Uranus passed in front of it. This indicated that rings around Uranus were blocking the light. The astronomers found nine of Uranus's rings this way.

CHAPTER THREE

STUDYING URANUS FROM SPACE

Voyager 2 is the only spacecraft to have flown by Uranus. It launched on August 20, 1977. Its twin, *Voyager 1*, took off days later on September 5, 1977. *Voyager 1* and *Voyager 2* were designed to study the outer planets. Both spacecraft flew by Jupiter and Saturn. Then they took different paths. After flying past Saturn, *Voyager 1* continued toward the edge of the solar system. In 2012, it became the

Voyager 2 was launched into space from Cape Canaveral, Florida.

UNITED STATES

This NASA illustration depicts a *Voyager* spacecraft traveling through space.

first spacecraft from Earth to leave the

solar system.

After Saturn, *Voyager 2* headed toward

Uranus and Neptune. It reached Uranus

in January 1986. The spacecraft did not

go into orbit around the planet. Instead,

Voyager 2 flew by the planet. At its closest point, *Voyager 2* was only 50,600 miles (81,500 km) away from Uranus's cloud tops.

Scientists did not know exactly what *Voyager 2* would see. Uranus's calm appearance came as a surprise. In *Voyager 2*'s pictures, Uranus had no bright cloud bands. It had no swirling storms. "We had been so spoiled by . . . the color . . . in the atmospheres of Jupiter and Saturn," says Carolyn Porco. She was a planetary scientist on the *Voyager 2* team. "Uranus was a little bit of a letdown."[5]

Scientists were not disappointed for long. *Voyager 2* discovered that Uranus is different from Jupiter and Saturn in many ways. Uranus is much smaller and colder. Different gases make up its atmosphere.

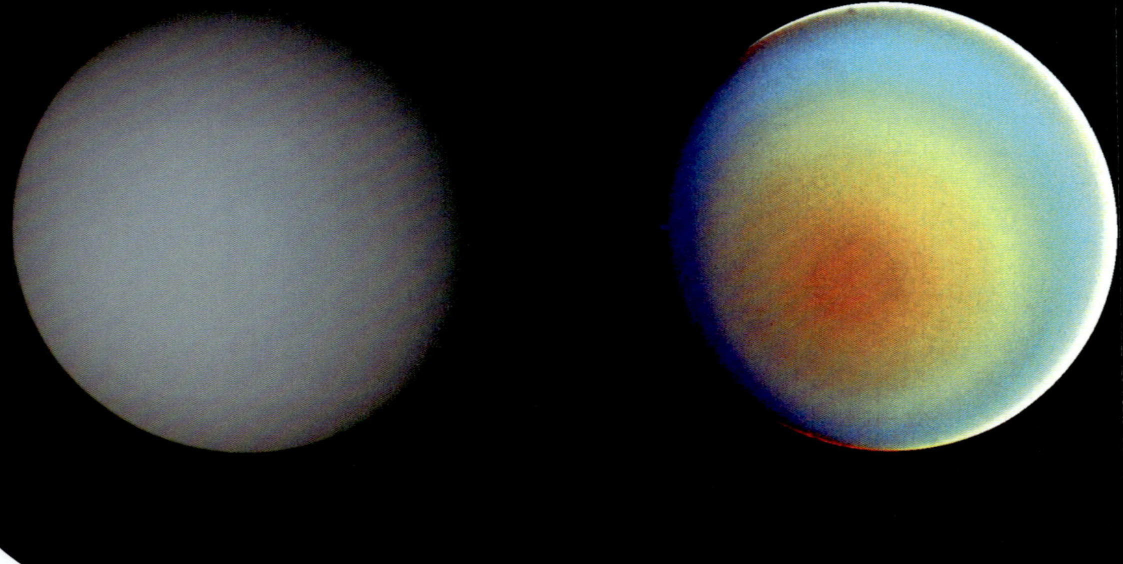

Long after *Voyager 2* flew by the planet, scientists kept studying the pictures and data the spacecraft collected. They learned lots of exciting new things. For example, *Voyager 2* discovered two new rings encircling the planet. They also learned more about Uranus's magnetic field. Scientists are still using data from *Voyager 2* today.

POWERFUL TELESCOPES

In 1990, the space shuttle *Discovery* carried five astronauts into orbit around Earth. Space shuttles were NASA spacecraft that could launch into space, return to Earth, and launch into space once again. The astronauts aboard *Discovery* placed a special telescope into orbit around Earth. The telescope was the Hubble

Close-Up Pictures of Uranus

Before *Voyager 2*, pictures of Uranus through telescopes revealed little about the planet. Engineers prepared the cameras on *Voyager 2* to take better pictures. They expected challenges. The sun's light at Uranus is dim. And the spacecraft would be moving fast. But everything went smoothly. *Voyager 2* successfully snapped close-up pictures of Uranus.

Space Telescope (HST). It would allow astronomers to see distant stars, comets, and planets clearly.

Telescopes on the Earth's surface must deal with **interference** from Earth's atmosphere. City lights affect telescope images as well. So, many large telescopes sit on mountaintops. The atmosphere is thinner at these great heights. And these locations are far away from cities. But even these telescopes deal with some interference.

Scientists built HST to avoid interference problems. It would orbit outside of Earth's atmosphere. The plan worked. HST has produced many beautiful images of the universe. It has also helped scientists learn more about planets, including Uranus.

HST powers itself with solar energy that it captures using two panels that resemble wings.

In 2005, HST took photographs of the distant planet. Astronomers looked at the pictures and discovered two new rings and two new moons. They named the small moons Mab and Cupid. One of the rings that Hubble discovered is twice the diameter of Uranus's other rings.

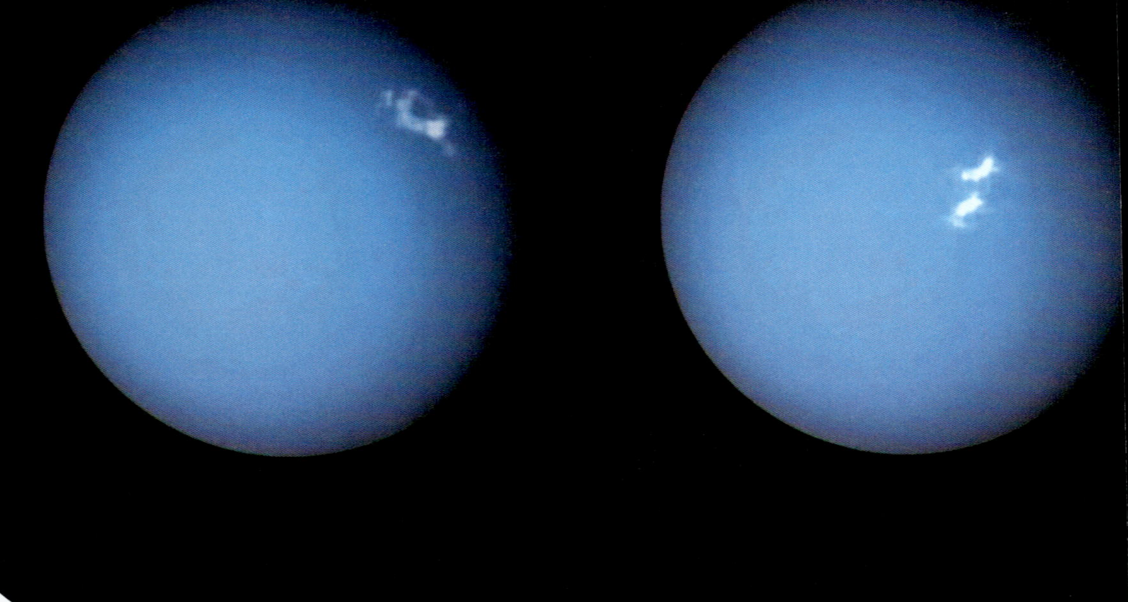

This HST image of Uranus revealed interactions between the Sun and the planet's magnetic field.

The James Webb Space Telescope (JWST) launched in 2021. It turned its camera toward Uranus in 2023. More studies of the planet are planned. Both HST and JWST will help scientists as they plan future missions to Uranus.

TODAY AND INTO THE FUTURE

Uranus fascinates scientists. This distant planet gives people a peek into the past. Scientists wonder where and how the ice giant formed. And they wonder what caused its axis to be so tilted. Studying Uranus helps people understand the early solar system. It shows them how planets change over time.

Uranus also teaches scientists about planets beyond the solar system. These worlds are called exoplanets. They orbit stars other than the Sun. Many exoplanets are Uranus's size. Some are made of similar materials. Studying the relatively close Uranus can help people better understand these exoplanets.

Voyager 2 flew by Uranus in 1986. But scientists keep learning from the data that the spacecraft collected. In 2024, scientists discovered something new. Days before the spacecraft had arrived at the planet, the planet's magnetic field was in an unusual state.

"If *Voyager 2* had arrived just a few days earlier, it would have observed a completely different magnetosphere at Uranus," says NASA scientist Jamie Jasinski.[6] A magnetosphere is a region of space affected by a planet's magnetic field. Astronomers want to send a spacecraft to Uranus that will stay for a while. They will compare the new data with *Voyager 2*'s. That way, they can observe how conditions around Uranus change over time.

Returning to Uranus is a major goal for scientists.

Returning to Uranus is important to astronomers. There is much more to learn about the seventh planet in the solar system. A long-term mission to Uranus would answer many questions. NASA is considering the idea of sending more spacecraft to the ice giant. One is planned to launch in the 2030s. It would orbit Uranus for many years.

THE MOONS OF URANUS

There are twenty-eight known moons of Uranus. The twenty-eighth moon was found in 2023. At first it was called S/2023 U1. Scientists planned to give it a permanent name later. Uranus's moons are sometimes called the literary moons. Their names come from characters in famous works of literature written by William Shakespeare and Alexander Pope.

This illustration of Uranus depicts some of the planet's moons orbiting among its rings.

Most of Uranus's moons are named after characters from plays by William Shakespeare.

William Herschel found Uranus's first two moons. More than 60 years passed before William Lassell discovered the next two. At first, Uranus's moons did not have names. Instead, they were numbered using Roman numerals from I to IV. Later, people began to call them Oberon, Titania, Umbriel,

and Ariel. In 1948, Gerard Kuiper found the
last of Uranus's five large moons. He named
it Miranda.

Voyager 2 found ten more moons when it
flew past Uranus. These small moons range
in size from 16 miles (26 km) to 96 miles
(154 km) across. Scientists continued
the literary tradition in naming these
new moons.

Astronomers have discovered more
moons since *Voyager 2* visited Uranus.
Some moons were found by studying
Voyager 2's data. Scientists used HST
to discover others. Astronomers using
powerful telescopes on Earth found even
more moons. In 2023, these telescopes
helped scientists discover three new moons
orbiting Uranus and Neptune. "The three

newly discovered moons are the faintest ever found around these two ice giant planets using ground-based telescopes," says astronomer Scott Sheppard.[7]

GROUPS OF MOONS

Uranus's moons can be divided into three main groups. The planet has thirteen inner moons. It has ten irregular moons. And it has five major moons.

The inner moons all orbit between Uranus and the major moon Miranda. Puck is the largest of the inner moons. It was the only inner moon photographed by *Voyager 2*. The inner moons are dark objects. They do not reflect much light. Cordelia and Ophelia are the moons closest to Uranus. These two moons are

THE LITERARY MOONS

Alexander Pope's
The Rape of the Lock

Ariel

Belinda

Umbriel

William Shakespeare's
As You Like It

Rosalind

William Shakespeare's
Hamlet

Ophelia

William Shakespeare's
King Lear

Cordelia

William Shakespeare's
The Merchant of Venice

Portia

William Shakespeare's
A Midsummer Night's Dream

Puck

Oberon

Titania

William Shakespeare's
Much Ado About Nothing

Margaret

William Shakespeare's
Othello

Desdemona

William Shakespeare's
Romeo and Juliet

Juliet

Mab

William Shakespeare's
The Taming of the Shrew

Bianca

William Shakespeare's
The Tempest

Ariel

Caliban

Ferdinand

Francisco

Miranda

Prospero

Setebos

Stephano

Sycorax

Trinculo

William Shakespeare's
Timon of Athens

Cupid

William Shakespeare's
Troilus and Cressida

Cressida

William Shakespeare's
The Winter's Tale

Perdita

The moon Ariel is named after characters sharing that name in works by both Shakespeare and Pope. This list does not include the unnamed twenty-eighth moon.

shepherd moons. This means they keep the material in one of Uranus's rings from drifting away.

Some of the inner moons are not stable. They pull and push on each other. In the future, the inner moons might cross orbits. They might collide. This could create a new ring around Uranus.

Irregular moons orbit planets in ways that astronomers consider unusual. For example, an irregular moon may have an

The Rings of Uranus

Uranus has thirteen rings. These rings are not bright like Saturn's. Uranus has dark, faint rings. The smallest may be less than 1 mile (1.6 km) wide. Uranus's rings do not have much dust in them. They are mostly made of small rocks.

oval-shaped orbit. Or it may orbit at a great
distance from the planet. For a long time,
astronomers thought that Uranus had no
irregular moons. But they had found such
moons orbiting every other giant planet in
the solar system. Finally, astronomers used
telescopes to spot Uranus's first irregular
moon in 1997. Since then, people have
found more irregular moons of Uranus.
They are generally very small. They range
in size from 5 miles (8 km) to 93 miles
(150 km) across.

THE MAJOR MOONS

The major moons Titania and Oberon
are the largest moons of Uranus by
diameter. Titania is the larger of the two. It
is about half the size of Earth's moon. It is

the eighth-largest moon in the solar system.
Titania is made of rock and ice. It might
have liquid water beneath its surface. It has

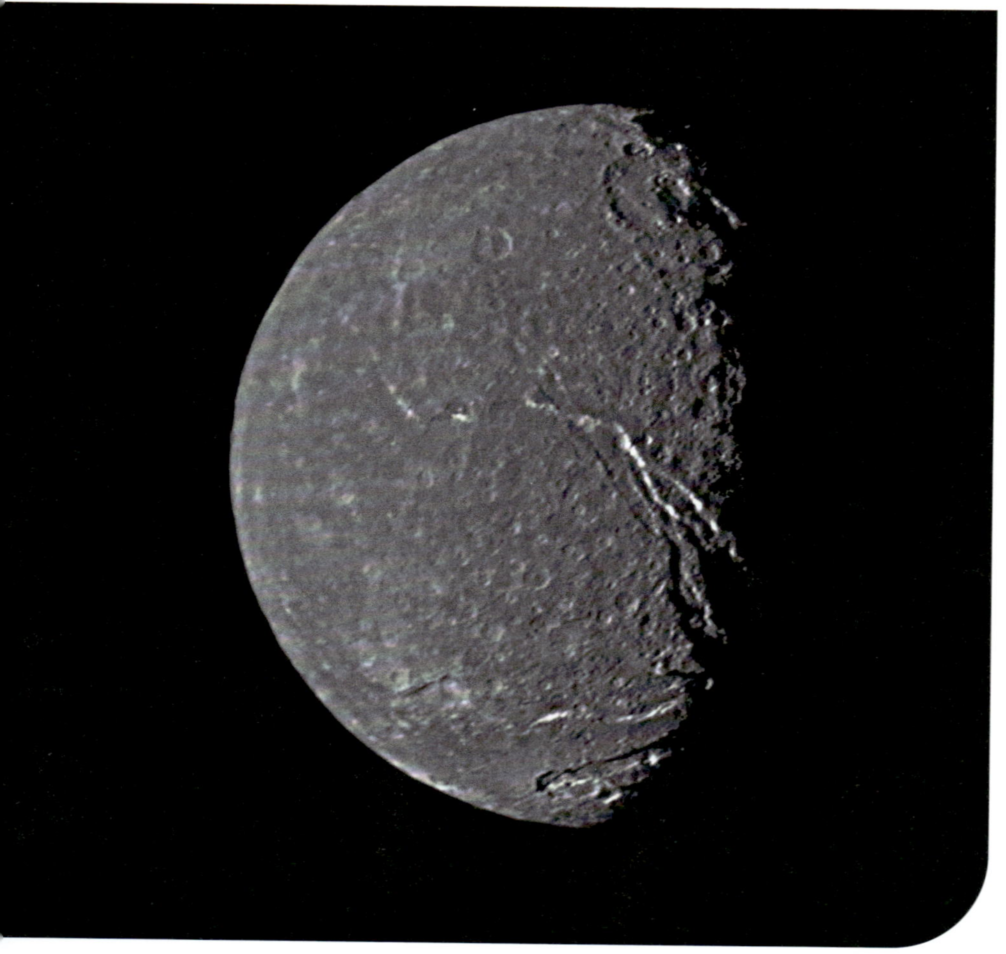

Voyager 2 was about 229,000 miles (369,000 km) away from Titania when it took this picture of the moon.

many **craters**. One of Titania's craters is 203 miles (326 km) across. Titania does not have an atmosphere. The moon's gravity is not strong enough to hold one in place.

Oberon is the outermost major moon of Uranus. It has more craters than Titania. It is also the reddest of Uranus's moons. Scientists think that Oberon has an icy surface covering a rocky core.

Umbriel is the darkest of the five large moons. It has many old, large craters. When *Voyager 2* flew past Umbriel, it took pictures of a bright circle about 90 miles (140 km) in diameter on the moon's surface. Scientists are not sure what caused this circle. It could be water ice around the edge of a crater.

Ariel has the brightest surface of all of Uranus's moons. It also seems to have

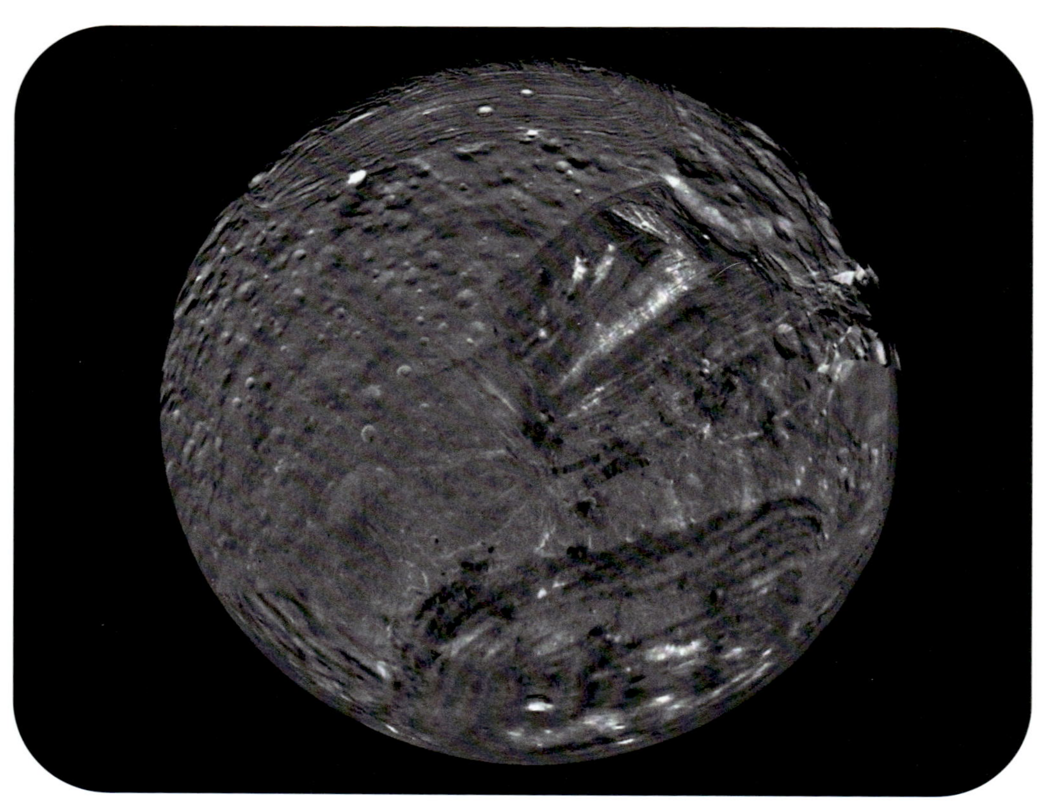

Miranda's surface is almost as bright as Ariel's.

the youngest surface. Most of the craters on Ariel are small. This suggests that recent collisions have erased older, larger craters. Valleys on its surface indicate past movement of large pieces of the moon's surface.

Miranda is the smallest of Uranus's five major moons. It orbits closest to the ice giant. It is mostly made of water ice and rock. *Voyager 2* flew close to this small moon. The spacecraft took many detailed pictures.

Miranda's surface is unique in the solar system. This moon is covered in giant canyons. One is twelve times deeper than Earth's Grand Canyon. Miranda also has craters, gorges, and cliffs. Scientists are fascinated by Miranda. They wonder how the small moon got its unusual surface.

GLOSSARY

astronomers
scientists who study space

atmosphere
gases that surround a planet

axis of rotation
an imaginary line running through a planet's poles around which the planet rotates

comet
a space object composed of ice and dust that leaves behind a long tail of material due to the Sun's energy

constellation
a group of stars forming a pattern in the night sky that people find significant

craters
bowl-shaped features in the ground usually caused by impacts or volcanoes

equator
an imaginary line drawn around a planet that is equally distant from both poles

interference
the act of blocking or disrupting something

SOURCE NOTES

CHAPTER ONE: AN ICE GIANT

1. Quoted in Jason Davis, "Uranus' Biggest Unsolved Mysteries," *The Planetary Society*, January 29, 2024. www.planetary.org.

2. Quoted in Davis, "Uranus' Biggest Unsolved Mysteries."

CHAPTER TWO: DISCOVERING URANUS

3. Quoted in J.L.E. Dreyer, *The Scientific Papers of Sir William Herschel*, Volume 1. The Royal Society and The Royal Astronomical Society, 1912, p. 100.

4. Mark Littmann, *Planets Beyond: Discovering the Outer Solar System*. John Wiley and Sons, 1988, p. 25.

CHAPTER THREE: STUDYING URANUS FROM SPACE

5. Quoted in "Giants," *The Planets*. BBC, 1999.

6. Quoted in "Mining Old Data from NASA's Voyager 2 Solves Several Uranus Mysteries," *Jet Propulsion Laboratory*, November 11, 2024. www.jpl.nasa.gov.

CHAPTER FOUR: THE MOONS OF URANUS

7. Quoted in Michael Roppolo, "Astronomers Discover 3 Tiny Moons Orbiting Uranus and Neptune," *CBS News*, February 29, 2024. www.cbsnews.com.

FOR FURTHER RESEARCH

BOOKS

KS Mitchell, *The Gas Giants: Jupiter, Saturn, Uranus, and Neptune.* BrightPoint Press, 2023.

Heather C. Morris, *Neptune.* BrightPoint Press, 2026.

Emily Schlesinger, *Astrobiology.* Saddleback Publishing, 2023.

INTERNET SOURCES

Conor McKeever, "Planet Uranus," *Natural History Museum*, n.d. www.nhm.ac.uk.

"Uranus Moons: Facts," *NASA*, January 27, 2025. https://science.nasa.gov.

"Uranus, the Sideways Planet," *The Planetary Society*, n.d. www.planetary.org.

Eyes on the Solar System
https://eyes.nasa.gov

This interactive website allows users to view a simulation of the solar system. It also offers basic facts about the planets and shows rings and the relative positions of moons.

The Schools' Observatory
www.schoolsobservatory.org

The Schools' Observatory is an online educational resource that guides students in observing the night sky. The website also has information for students interested in pursuing a career in astronomy.

Webb Space Telescope
https://webbtelescope.org

The James Webb Space Telescope allows astronomers to view distant parts of the universe up close. This website collects some of the best of the telescope's images. Visitors can search for specific images using a keyword search option.

Cover: © NASA

5: © NASA

7: © Francis Reddy/NASA

8: © Joseph DePasquale/NASA

10: © NASA

13: © Hakan Akirmak Visuals/Shutterstock Images

14: © NASA

17: © Kevin M. Gill/NASA

18: © NASA

21: © Lawrence Sromovsky/NASA

22: © NASA

25: © Syafiqqmn/Shutterstock Images

26: © T. Gibson/Smithsonian Libraries

28: © Procy/Shutterstock Images

29: © Emery Walker/Herschel

33: © NASA

35: © NASA

36: © NASA

38: © NASA

41: © NASA

42: © NASA

45: © Julian Baum/Science Source

47: © Claus Lunau/Science Source

48: © Stocksnapper/Shutterstock Images

51: © Red Line Editorial

54: © NASA

56: © NASA

ABOUT THE AUTHOR

Heather C. Morris writes books for students who love science. Before writing full time, she worked as a scientist for a NASA contractor, talking to astronauts on the International Space Station and testing new equipment in extreme environments.